A KEY TO THE

BRITISH AND EUROPEAN FRESHWATER BRYOZOANS

by

S. P. MUNDY

Department of Zoology
University College of Swansea

FRESHWATER BIOLOGICAL ASSOCIATION
SCIENTIFIC PUBLICATION No. 41

1980

PREFACE

Groups of animals with small numbers of freshwater representatives that are not readily apparent to the casual observer have tended to be neglected by biologists. One such group is the Bryozoa, so we are fortunate that Dr Mundy has been prepared to write this introductory guide and key. We hope that its availability will stimulate the study of this group. Species of Bryozoa have already been shown to play a significant role in the economy of some freshwater ecosystems (e.g. *Plumatella* in the River Thames at Reading), and this too should encourage further research. Dr Mundy would be very grateful for information on the distribution of freshwater bryozoans particularly if accompanied by specimens of colonies or statoblasts*.

The Ferry House — E. D. Le Cren
February 1980 — *Director*

* Dr Mundy's current address is: The Marine Laboratory, Ferry Road, Hayling Island PO11 0DG.

ISBN 0 900386 39 8
ISSN 0367–1887

CONTENTS

INTRODUCTION

Despite the early discovery of *Lophopus crystallinus* by Trembley in 1744, and the publication of a beautifully illustrated monograph by Allman in 1856, the study of freshwater bryozoans is still in its infancy. Some possible reasons for the neglect of this interesting group of animals are discussed below. The study of marine bryozoans has followed a similar course, although the founding of the International Bryozoological Association in 1965 reflects a recent upsurge of interest.

The phylum Bryozoa (once known as the Polyzoa) can be divided into three distinct classes: the Gymnolaemata, the Stenolaemata and the Phylactolaemata. The Gymnolaemata contains a small number of freshwater species, the rest being marine along with the Stenolaemata. Both groups together comprise about 4,000 living species (Ryland 1970). The Phylactolaemata, which is confined to fresh water, contains far fewer species, but these are often common and widely distributed. Of about 35 recognized species in 8 genera, 12 occur in Europe and 9 in the British Isles. Several of these are believed to have world-wide distributions.

Existing knowledge of the Phylactolaemata is based on the work of a small number of authors, mostly in Europe. Allman (1850, 1856) provided the first systematic account of the group, and Kraepelin (1887) and Braem (1890) studied the histology of several species. Marcus (1925, 1926, 1934, 1940), Cori (1941), Rogick (1934, 1935, 1959), Toriumi (1941, 1956, 1974), Prenant & Bobin (1956), Wiebach (1958), Hyman (1959) and Brien (1960) later added to the histological background and repeatedly reviewed and amended the taxonomy. Since Allman (1879), only Harmer (1901, 1913, 1931) has made a significant contribution to the study of British species. More recently, Bushnell (1965–1974) studied the ecology of some American species, Lacourt (1968) proposed another taxonomic revision, and Mukai (1973, 1974) and Tajima & Mukai (1975) have made various histological observations.

A major difficulty in the past has been the identification of certain species. For instance, the extensive work of Dr A. B. Hastings (see Ryland 1970) on British freshwater bryozoans has remained unpublished because of taxonomic uncertainties. Since there are so few phylactolaemate species, identification is easy in most cases, but existing keys are all unsatisfactory in some degree. The taxonomy of the principal genus, *Plumatella* Lamarck 1816, is particularly confused, largely because some taxonomists have placed undue emphasis on a few very variable morphological characters such as the colour or length of zooids, the number of tentacles, or the shape and size of statoblasts. An objective assessment of the status of the more problematic species of *Plumatella* has been provided by the use of electrophoretic techniques supported by stereoscan electron microscopy (Mundy 1979; Mundy & Thorpe 1979; in press; Thorpe & Mundy, in press), and the results of this work are followed here.

Phylactolaemates are very common, being found in a wide variety of freshwater habitats throughout the world. Nevertheless, they can be difficult to find, as they have a rather short growing season as adult colonies, at least in temperate climates. Because they are sessile, they cannot easily be collected in a net unless a fine mesh is used to obtain the resting bodies known as statoblasts, and experimental work is hampered by the difficulty of culturing colonies in the laboratory. These problems provide some explanation of the scarcity of published information on the ecology of the phylactolaemates, the only detailed work being that of Bushnell (1965a, b, c; 1966; 1968).

The traditional view of their phylogeny suggests that phylactolaemates are primitive and were ancestral to the marine bryozoans. However, since they are poorly represented in the fossil record, except for statoblasts of a few species which have been found in Tertiary and more recent deposits, any phylogenetic scheme is necessarily tentative. In particular, the relationship between marine bryozoans, phylactolaemates and the Phoronida is not clear, and the so-called primitive status of phylactolaemates is not accepted by some workers, including the present author.

Only two gymnolaemate bryozoans are found in fresh waters in Britain and Europe. Both are members of Ctenostomata, and even less is known of their biology than of the phylactolaemates. Their taxonomy has been reviewed by Toriumi (1952), Prenant & Bobin (1956) and Hyman (1959).

GENERAL CHARACTERS

The Phylactolaemata are small colonial invertebrates (individuals are known as *zooids*) with a retractile crown (*lophophore*) of ciliated *tentacles* surrounding the *mouth*, which protrudes through the *orifice* and is used in feeding (see fig. 1). The gut is recurved, bringing the *anus* near to the mouth but outside the lophophore. The colonies are sessile, with the exception of *Cristatella mucedo* which is capable of slow movement of the

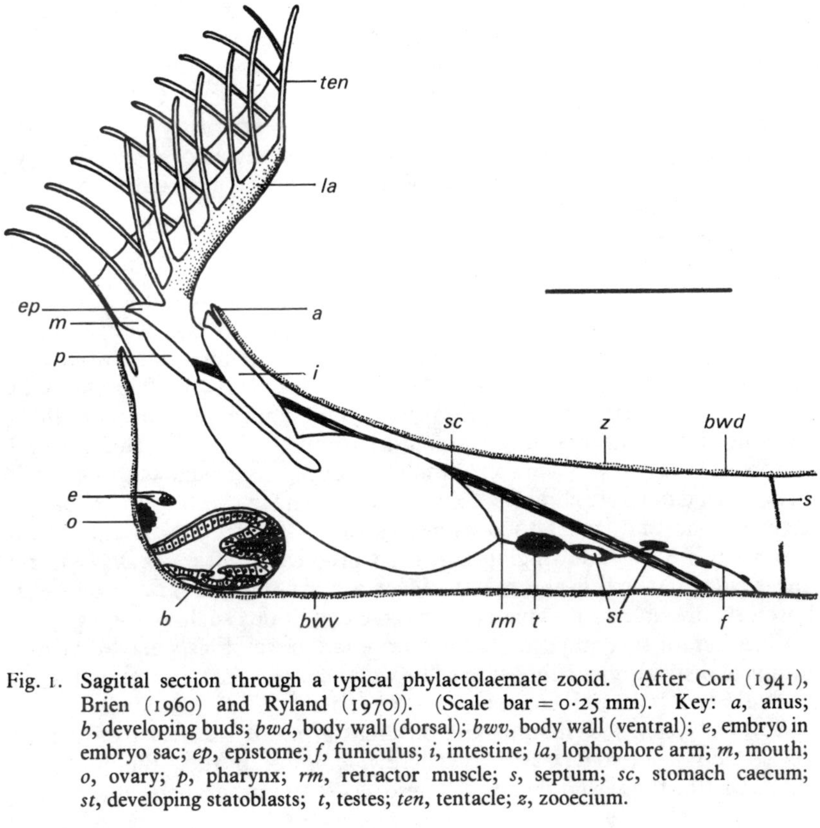

Fig. 1. Sagittal section through a typical phylactolaemate zooid. (After Cori (1941), Brien (1960) and Ryland (1970)). (Scale bar = 0·25 mm). Key: *a*, anus; *b*, developing buds; *bwd*, body wall (dorsal); *bwv*, body wall (ventral); *e*, embryo in embryo sac; *ep*, epistome; *f*, funiculus; *i*, intestine; *la*, lophophore arm; *m*, mouth; *o*, ovary; *p*, pharynx; *rm*, retractor muscle; *s*, septum; *sc*, stomach caecum; *st*, developing statoblasts; *t*, testes; *ten*, tentacle; *z*, zooecium.

order of a few centimetres in a day. A flap of tissue, the *epistome*, overhangs the mouth, and has been regarded as a distinct body region known as the protosome, with the lophophore representing the mesosome and the rest of the body of the metasome. However, in the absence of clear embryological evidence, this is a rather arbitrary division since all three parts are confluent, allowing coelomic fluid to circulate freely. There is no vascular system, oxygen being obtained from the surrounding water by simple diffusion. The lophophore is U-shaped in all species except the members of the family Fredericellidae, in which it is oval. Colonies range in size from less than a centimetre to several metres across.

Asexual reproduction is an important feature of phylactolaemate biology, and is accomplished in two ways. Daughter zooids are budded from the ventral body wall (see fig. 1), thereby producing a colony, and although a partial *septum* may be present between zooids, the coelomic fluid is free to pass from one zooid to another. In mature zooids, overwintering bodies known as *statoblasts* are formed on a cord of tissue called the *funiculus* which passes from the outer curve of the gut to the ventral body wall. Statoblasts consist of two chitinous valves forming the *capsule* (see fig. 2) and enclosing a yolk mass. Most statoblasts have another chitinous layer, the *periblast*, which overlies the capsule and may form a gas-filled *annulus* which allows the statoblast to float (fig. 2). Non-floating statoblasts may have a smaller annulus, reduced to a ridge (e.g. fig. 10C; 11D). In temperate climates, colonies degenerate in the autumn, releasing between one and 30 statoblasts per zooid. Many of these statoblasts germinate in the following spring, each producing an *ancestrula* (plural *ancestrulae*), which is the first zooid of a new colony. The statoblasts, however, are able to survive very adverse conditions such as freezing and desiccation for several years. Before germination statoblasts may serve as a means of distribution, being carried by water currents or wind, or on the fur or feathers of waterside animals.

Several systems of naming the different types of statoblast have been devised, but just two terms are used here; *floatoblast* for those that float, and *sessoblast* for those that do not. These terms are rather inelegant, but are

Fig. 2. Statoblasts.
A, *Fredericella australiensis* (sessoblast); B, *F. sultana* (sessoblast); C, *Lophopus crystallinus* (floatoblast); D, *Pectinatella magnifica* (floatoblast); E, *Cristatella mucedo* (floatoblast, dorsal face); F, *Plumatella fruticosa* (floatoblast); G, *P. casmiana* (thick-walled floatoblast); H, *P. emarginata* (floatoblast, ventral face); I, *P. emarginata* (floatoblast, dorsal face); J, *Hyalinella punctata* (floatoblast); K, *Plumatella repens* (floatoblast); L, *P. coralloides/fungosa* (floatoblast).
(Scale bars = 300 μm.)

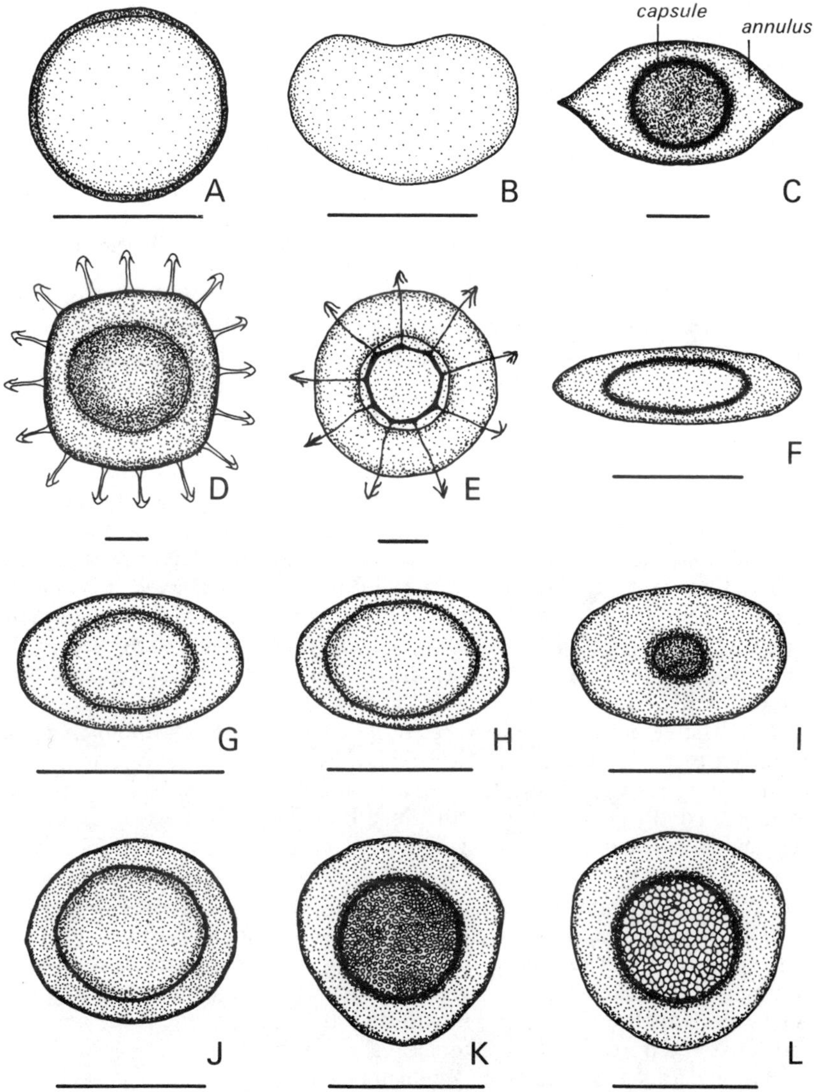
capsule
annulus
A
B
C
D
E
F
G
H
I
J
K
L

widely used and are useful for distinguishing the two types when they occur together in a single species. It is not intended that any natural division of function should be implied, but it does seem likely that the presence of a gas-filled annulus in floatoblasts represents an important functional advance. Detailed statoblast morphology is usually species-specific, and is in its simplest form in the sessoblasts of the Fredericellidae. The Plumatellidae possess both sessoblasts and floatoblasts, whilst the Lophopodidae and Cristatellidae have only floatoblasts.

Sexual reproduction involves the production of eggs and sperm, each zooid being hermaphroditic. As in many freshwater animals, embryology and larval development are highly modified. The embryo develops in a ventral *embryo sac*, resulting in a juvenile colony rather than a true larva. A ciliated mantle encloses the "larva", and allows it to swim briefly until it reaches a substratum suitable for attachment and development. Larvae are released from May to August.

The diet of freshwater bryozoans consists of minute organisms such as protozoans, bacteria, rotifers, diatoms, and occasionally small crustaceans. Cilia borne on the tentacles create a current of water through the lophophore, and selected particles are directed towards the mouth.

The body wall consists of several distinct layers. The innermost is a ciliated peritoneum, overlain by a longitudinal muscle layer, a basement membrane (which may be fibrous in certain parts of some gelatinous species), and a circular muscle layer. The muscle layers are used in the protrusion of the lophophore, withdrawal being effected very rapidly by a powerful *rectractor muscle* (see fig. 1). The outermost layer of the body wall is the epidermis which secretes the chitinous or gelatinous material referred to in the key as the *zooecium*. Despite its archaic connotation, which suggests the ill-conceived notion of an animal living in a box (see Ryland, 1970), this term is useful since it denotes that part of the colony which may remain intact when zooids degenerate, often with statoblasts still inside. All parts of the zooid apart from the zooecium are known collectively as the *polypide*.

The nervous system consists of a *ganglion* situated between the anus and the epistome, with branched nerves passing down each arm of the lophophore and into the tentacles. The epistome is well-innervated from a ring of nerves which encircles the mouth and also supplies the three tentacles adjacent and dorsal to the epistome.

More detailed accounts of phylactolaemate biology are provided by Prenant & Bobin (1956), Wiebach (1958), Hyman (1959) and Brien (1960). Of these, only that of Hyman is in English. Ryland (1970) has given a good introduction to most aspects of freshwater and marine bryozoology for biologists and geologists.

The Gymnolaemata are represented by two ctenostome species, *Paludicella articulata* and *Victorella pavida*. The latter is strictly speaking a brackish water species, tolerating both fresh water and sea water, whilst *P. articulata* is wholly freshwater in distribution.

Both species differ from the phylactolaemates in a number of important ways. For instance, the lophophore is circular instead of U-shaped, and they lack both an epistome and statoblasts. However, they do produce over-wintering bodies known as *hibernacula* (see fig. 6A) which are modified zooids with a slightly thickened chitinous exterior. They are analogous to statoblasts in function only, consisting of a single, variously-shaped outer layer enclosing a dormant zooid, rather than the typical bivalve statoblast structure.

For further details of the biology of these and other gymnolaemates, the reader is referrred to Prenant & Bobin (1956, 1966), Hyman (1959) and Ryland (1970, 1976). Ryland & Hayward (1977) and Hayward & Ryland (1979) have recently produced excellent accounts of the British marine Cheilostomata, including keys and very readable summaries of the biology of marine bryozoans.

COLLECTION AND PRESERVATION

Colonies may be found from March to mid-November in almost every lake and in many ponds, streams and rivers, unless heavy pollution or silting is present. Emergent plants such as water-lilies and reeds are common substrata, as are submerged stones and wood. Fredericellids can be difficult to find as they tend to live at the bottom of the water column, and some form of grab or hook may be useful. To the naked eye, the Fredericellidae and Plumatellidae have the appearance of branching brown tubes forming encrusting, cushion-like or bushy colonies. They may be very difficult to see on dark substrata such as rocks and bark. Gelatinous forms (Lophopodidae, Cristatellidae) are usually pale yellow in colour with statoblasts or gut contents showing as dark patches. *Cristatella* and *Lophopus*, and to a lesser extent *Pectinatella*, tend to collapse when removed from the water, and are most easily spotted whilst still submerged. Many freshwater bryozoans may, at a first glance, be confused with mosses or the roots of plants, but a closer look, preferably with the assistance of a hand-lens, soon reveals the presence of statoblasts, the lophophore, and the characteristic U-shaped gut. Other animals such as *Hydra* spp. or sponges have neither the stiffly-held ciliated tentacles nor the branching growth habit of bryozoans. *Cristatella* and *Lophopus* may sometimes be mistaken for the egg-masses of molluscs, but again a careful examination soon reveals their internal structure. Colonies are best examined whilst still alive, using a hand-lens or low-power microscope. They will survive for a considerable time if kept in several centimetres of the water in which they are found, with silt excluded and with gentle aeration if possible. Ideally, they should be left intact on their substrata, but otherwise may be carefully lifted free with a sharp blade.

If immediate fixation is necessary, it is useful, for the first part of the key, to note first whether the lophophore is U-shaped or circular, but this is not essential with mature colonies containing many statoblasts. Suitable preservatives are 70% alchohol or 4% formalin. Narcotization requires the gradual addition of a few magnesium chloride or menthol crystals over several hours, during which the colonies should be left completely undisturbed.

Floatoblasts, and occasionally sessoblasts, occur in the surface drift material of ponds and lakes throughout the year, and may be collected by skimming the surface with a fine net (< 250 µm mesh). They may be mounted in most standard preparations such as canada balsam, after dehydration through graded alcohols, which may be fairly rapid (10 min). Certain features are more easily and reliably observed if the statoblasts are first split, as described for some *Plumatella* species in the footnote on page 22.

CLASSIFICATION AND CHECKLIST

The British and European Phylactolaemata comprises four families with two distinct colony forms; the branching, chitinous colonies of the Fredericellidae (2 species in one genus) and Plumatellidae (7 species in two genera), and the gelatinous colonies of the Lophopodidae (2 species in two genera) and Cristatellidae (one species). The Fredericellidae are simplest in form, having a circular or oval lophophore and simple sessoblasts. The Plumatellidae are very diverse and can be difficult to identify, but the species of interest here are fairly distinct, particularly if floatoblasts as well as colonies are available. The Lophopodidae and Cristatellidae have well-defined colony forms, and floatoblasts with characteristic spines and hooks. *Cristatella mucedo* in particular has a complex floatoblast, and with its mobile colonies possesses the most complicated morphology of any member of the Phylactolaemata.

Paludicella articulata and *Victorella pavida* are members of the order Ctenostomata within the class Gymnolaemata. A summary of the classification of marine bryozoans is given by Ryland (1970).

Class PHYLACTOLAEMATA Allman 1856

Family FREDERICELLIDAE Hyatt 1868

Genus FREDERICELLA Gervais 1838

Fredericella australiensis Goddard 1909

F. sultana (Blumenbach 1779)

Family PLUMATELLIDAE Allman 1856

Genus HYALINELLA Jullien 1885

Hyalinella punctata (Hancock 1850)

Genus PLUMATELLA Lamarck 1816

Plumatella casmiana Oka 1907

P. coralloides Allman 1850

P. emarginata Allman 1844

P. fruticosa Allman 1844

P. fungosa (Pallas 1768)

P. repens (Linnaeus 1758)

Family LOPHOPODIDAE Rogick 1935

Genus LOPHOPUS Dumortier 1835

Lophopus crystallinus (Pallas 1768)

Genus PECTINATELLA Leidy 1851

Pectinatella magnifica (Leidy 1851)

Family CRISTATELLIDAE Allman 1856

Genus CRISTATELLA Cuvier 1798

Cristatella mucedo Cuvier 1798

Class GYMNOLAEMATA Allman 1856

Order CTENOSTOMATA Busk 1852

Genus PALUDICELLA Gervais 1836

Paludicella articulata (Ehrenberg 1831)

Genus VICTORELLA Saville Kent 1870

Victorella pavida Saville Kent 1870

THE KEY

1 Zooecium soft, gelatinous and transparent, may be lobed but never branching; no distinct zooecial tubes (figs 3; 4A; 5A)— **2**

— Zooecium chitinous and branching; distinct zooecial tubes (figs 6; 7A, B; 8; 9A, B; 10A; 11A, B) **4**

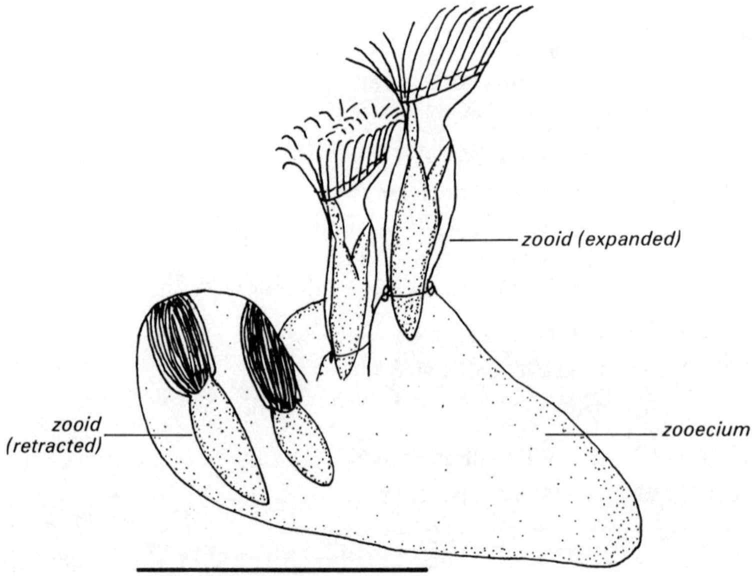

Fig. 3. Colony form. *Lophopus crystallinus* (scale bar = 1 mm).

2 Floatoblasts produced into acute points at the poles (fig. 2C); zooids all directed towards one end of sac-like or lobed colony; up to 70 tentacles (fig. 3)— **Lophopus crystallinus** (Pallas)

— Floatoblasts with hooks around the margin (fig. 2D, E), never with acute points at the poles; zooids arranged in rosette patterns (fig. 4B), parallel rows (fig. 5A), or concentric circles— **3**

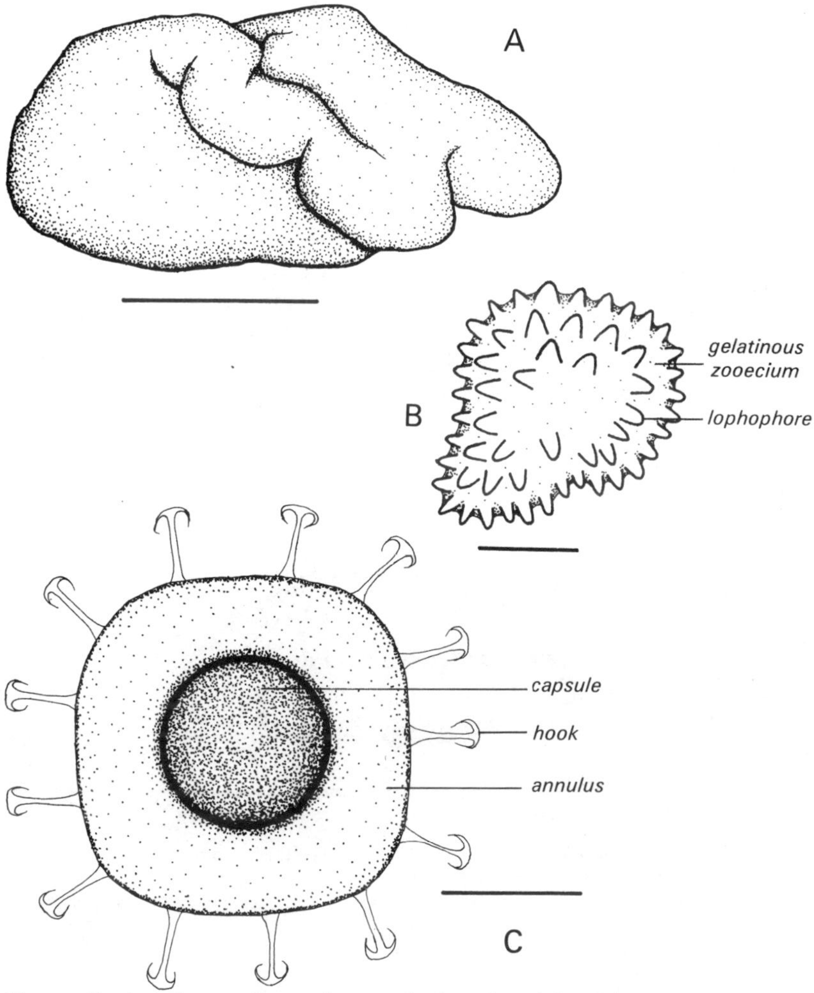

Fig. 4. *Pectinatella magnifica.* A, general colony form (after Oda, 1974) (scale bar = 0·5 m); B, single rosette of colony (scale bar = 10 mm); C, floatoblast (scale bar = 500 µm).

3 Floatoblasts curved about 2 axes (saddle-shaped), with a single row of flattened, hooked spines arising from the outer edge of the annulus (figs 2D; 4C); colonies often large (about 2 m across), zooids arranged in rosette patterns on the surface (fig. 4A, B); up to 90 tentacles—
Pectinatella magnifica (Leidy)

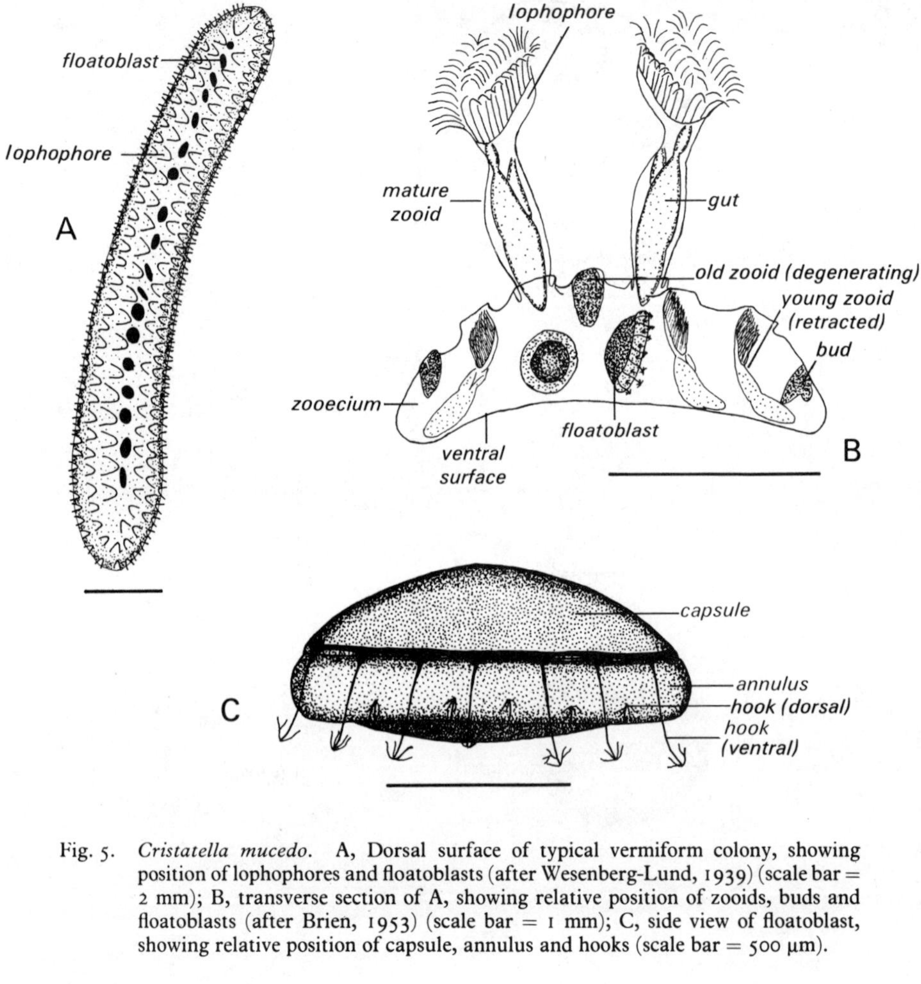

Fig. 5. *Cristatella mucedo.* A, Dorsal surface of typical vermiform colony, showing position of lophophores and floatoblasts (after Wesenberg-Lund, 1939) (scale bar = 2 mm); B, transverse section of A, showing relative position of zooids, buds and floatoblasts (after Brien, 1953) (scale bar = 1 mm); C, side view of floatoblast, showing relative position of capsule, annulus and hooks (scale bar = 500 µm).

— Floatoblasts perfectly circular, the annulus attached only to the dorsal valve, with 2 rows of hooked cylindrical spines, one dorsal and one ventral, arising from the capsule (figs 2E; 5C); colony vermiform (1·5–5·0 mm wide; up to 400 mm, but usually less than 100 mm long), zooids arranged in parallel rows (fig. 5A, B) or, in small colonies, in concentric circles; up to 100 tentacles— **Cristatella mucedo** Cuvier

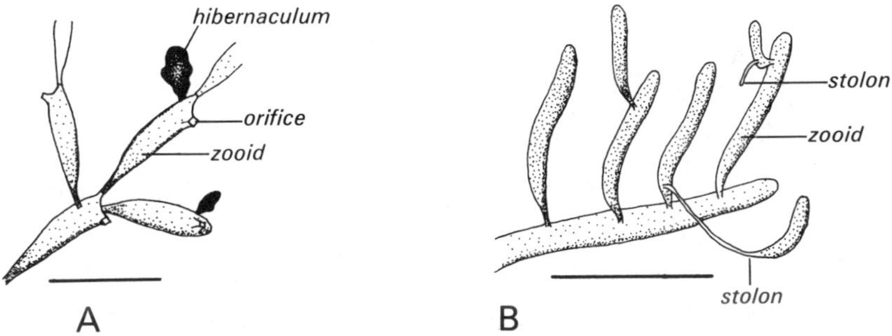

Fig. 6. Colony forms. A, *Paludicella articulata* (scale bar = 1 mm); B, *Victorella pavida* (scale bar = 1 mm).

4(1) No statoblasts produced; epistome absent; colony consisting of claviform tubes (fig. 6A) or a creeping stolon with zooids arising from it (fig. 6B)— **5**

— Statoblasts produced; epistome present; colony consisting of cylindrical branching tubes (e.g. figs 7A, B; 8A, B; 9A, B; 10A; 11B)— **6**

5 Colony consisting of claviform tubes (fig. 6A); 10–20 tentacles; in fresh water only— **Paludicella articulata** (Ehrenberg)

— Colony consisting of a creeping stolon with zooids arising from it (fig. 6B); 8 tentacles; usually in brackish water— **Victorella pavida** Saville Kent

6(4) Lophophore circular or nearly so; all statoblasts of sessoblast type (no floatoblasts)— **7**

— Lophophore U-shaped; sessoblasts and floatoblasts produced— **8**

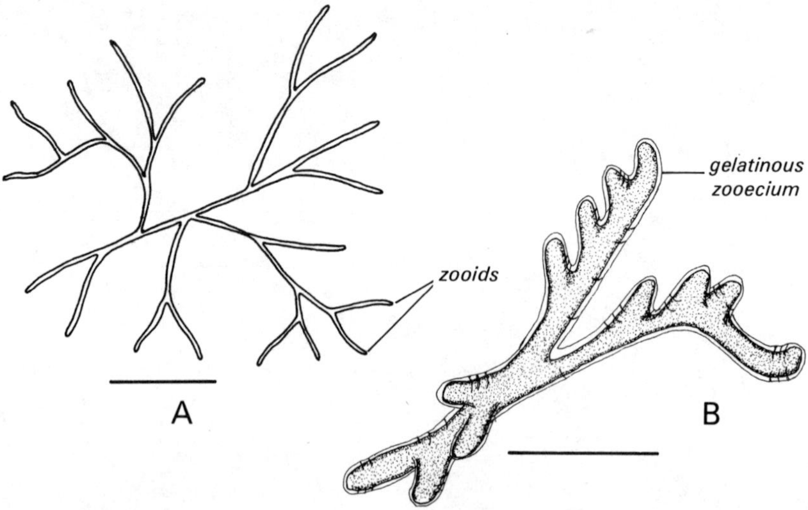

Fig. 7. Colony forms. A, *Fredericella* spp. (scale bar = 2 mm); B, *Hyalinella punctata* (scale bar = 5 mm).

7 Sessoblasts circular or nearly so, with a distinct annulus (fig. 2A); up to 35 tentacles— **Fredericella australiensis** Goddard

— Sessoblasts oval or reniform, with little or no annulus (fig. 2B); up to 30 tentacles— **Fredericella sultana** (Blumenbach)

8(6) Zooecium thick, soft and translucent (fig. 7B); floatoblasts and sessoblasts circular (fig. 2J); up to 70 tentacles— **Hyalinella punctata** (Hancock)

— Zooecium thin, various shades of brown; floatoblasts oval or elongate (fig. 2F, G, H, I, K, L); sessoblasts oval (figs 10C; 11D)— **9**

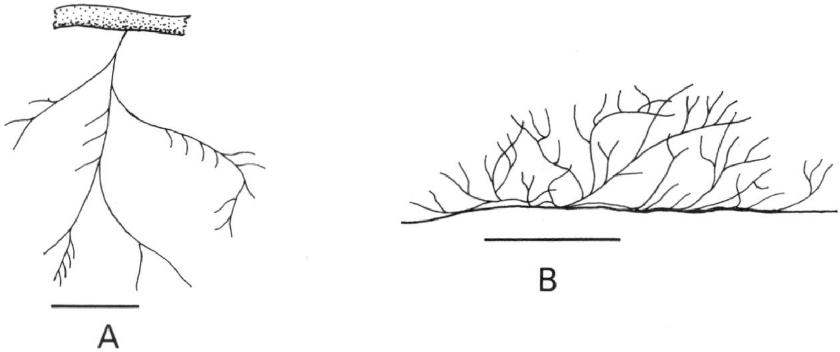

Fig. 8. Colony forms. A, *Plumatella fruticosa* (scale bar = 5 mm); B, *Plumatella coralloides* (scale bar = 10 mm).

9 Floatoblasts more than 2·5 times as long as broad; annulus much wider at the poles than laterally (fig. 2F); colonies with openly branching, mostly erect zooecia (fig. 8A); free zooecia usually triangular in cross-section; up to 60 tentacles— **Plumatella fruticosa** Allman

— Floatoblasts less than 2·5 times as long as broad, annulus almost the same width all round (fig. 2G, H, I, K, L); colonies mostly encrusting, (figs 9A, B; 10A), or forming spongy or bushy masses (figs 8B; 11A, B). Free zooecia may be keeled but are rarely triangular in cross-section— **10**

10 Floatoblast capsule covered by annulus much more dorsally than ventrally (fig. 2H, I); floatoblast flattened dorsally; up to 60 tentacles— **Plumatella emarginata** Allman

— Floatoblast capsule covered by annulus only slightly more dorsally than ventrally; floatoblast valves equiconvex or nearly so— **11**

11 Two floatoblast types: a common elongate type (up to twice as long as broad) with a thin soft capsule (fig. 9C), and a rarer type (fig. 2G) which is broadly oval with a thick capsule; up to 50 tentacles— **Plumatella casmiana** Oka

— Floatoblasts of one type only: broadly oval (often a rounded triangle) with a thick capsule (figs 2K, L; 10B; 11C)— **12**

Fig. 9. A, *Plumatella emarginata*: colony (scale bar = 1 mm); B, *Plumatella casmiana*: colony (scale bar = 1 mm); C, *Plumatella casmiana*: thin-walled floatoblast (scale bar = 100 μm).

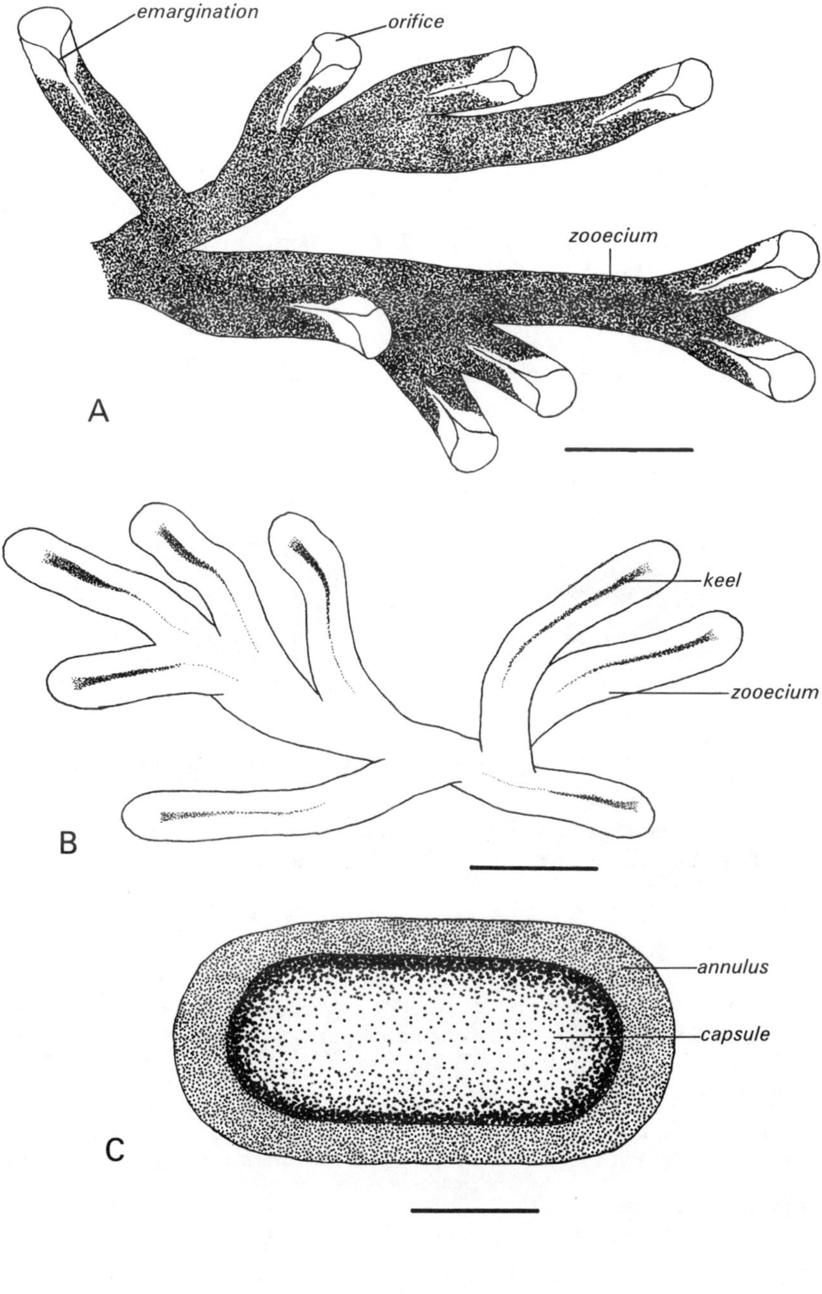
emargination
orifice
zooecium
A
keel
zooecium
B
annulus
capsule
C

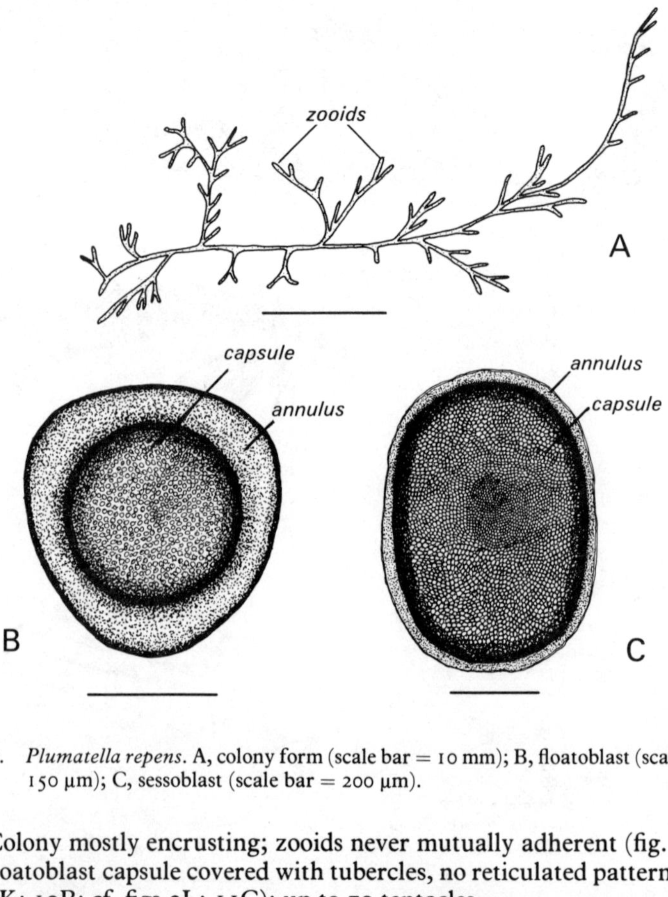

Fig. 10. *Plumatella repens*. A, colony form (scale bar = 10 mm); B, floatoblast (scale bar = 150 μm); C, sessoblast (scale bar = 200 μm).

12 Colony mostly encrusting; zooids never mutually adherent (fig. 10A); floatoblast capsule covered with tubercles, no reticulated pattern* (figs 2K; 10B; cf. figs 2L; 11C); up to 70 tentacles—
Plumatella repens (Linnaeus)

— Colony partly erect; zooids mutually adherent for all or part of their length (figs 8B; 11A, B); floatoblast capsule (and sometimes annulus) covered with ridges in a reticulated pattern,* often with interstitial tubercles (fig. 2L; cf. fig. 2K)— **13**

* The surface patterns on the capsule can be accurately observed with a light microscope only if the floatoblast is first split into two valves. This is best done by heating some floatoblasts in 10% potassium hydroxide (KOH) for a few minutes, teasing the valves apart, and mounting them on a slide. Identifications made on the basis of these patterns should be supported by reference to colony growth forms whenever possible.

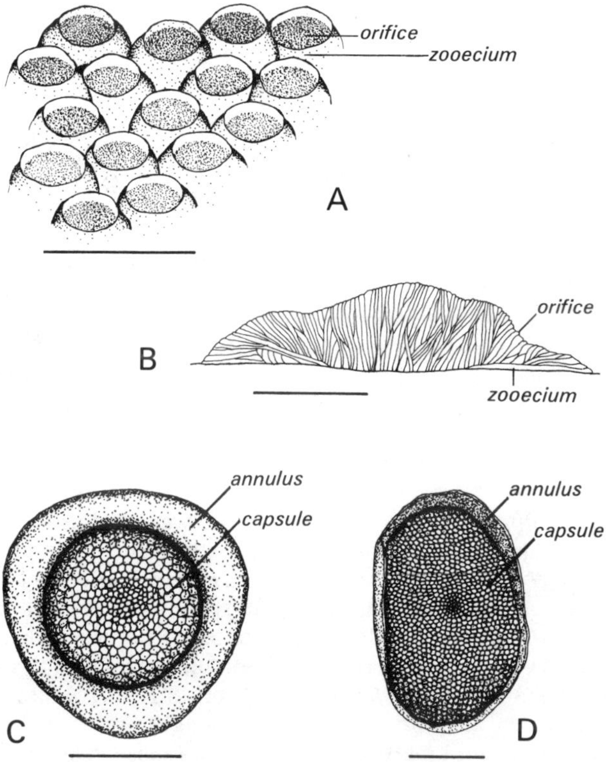

Fig. 11. *Plumatella fungosa*. A, surface of colony (scale bar = 1 mm); B, transverse section of colony (scale bar = 10 mm); C, floatoblast (scale bar = 150 µm); D, sessoblast (scale bar = 200 µm).

13 Zooids mutually adherent throughout their length, forming dense spongy colonies with a honeycomb-like appearance on the surface (fig. 11A, B); up to 70 tentacles— **Plumatella fungosa** (Pallas)

— Zooids mutually adherent to some extent, but never throughout their length; colonies consisting of long intertwining zooids with a bushy appearance (fig. 8B); up to 70 tentacles— **Plumatella coralloides** Allman

NOTES ON SPECIES

These notes are intended to help confirm identification, and to provide additional information. Little is known of the relative abundance of species, and this aspect is mentioned only when a species is notably rare or common. All species occur throughout Britain and Europe except when otherwise indicated.

Few clear relationships between habitat type (water chemistry, biological status, etc.) and the occurrence of species are apparent, except that *Cristatella mucedo* is found most often in eutrophic conditions whilst *Fredericella sultana* prefers oligotrophic waters. *Plumatella fruticosa* and *Pectinatella magnifica* are most abundant in still water, whilst other species are frequently found in rivers and streams, particularly at the outflow of a lake or pond. *Hyalinella punctata*, *Lophopus crystallinus* and *Plumatella coralloides* are by far the least common species on present evidence, *Lophopus crystallinus* apparently being confined to the east of England.

Sites in Britain where each species has been found are given in small type and National Grid References are provided for all recent records. Some species are probably ubiquitous, but it was thought preferable to provide only specific records until more detailed information is available.

Cristatella mucedo (figs 2E; 5)

Colonies are round when small, becoming elongate as they grow larger. Large colonies may divide into several smaller colonies. The zooids are arranged in parallel or concentric rows, the more mature ones innermost and developing ones in successive rows towards the edge (fig. 5A, B). Floatoblasts accumulate in the middle of the colony. The lophophores are relatively unreactive to stimulation, and rarely retract rapidly in the way that other species do when disturbed. Colonies are usually pale yellow in colour, and are found on most substrata in still or running water.

Alderfen Broad, Norfolk (TG355194); Blagdon Lake, Avon (ST508608); Chew Lake, Avon (ST572615); Galway, W. Ireland; Grasmere, Cumbria (NY343061), Margam, W. Glamorgan (SS8086); Quoisley Meres, Cheshire (SJ547455); River Rothay, Cumbria (NY347064); Upton Broad, Norfolk (TG388134); Wrayesbury Reservoir, Berkshire (TQ0274); River Yare, Norfolk.

Fredericella australiensis (figs 2A; 7A) and **F. sultana** (figs 2B; 7A)

The validity of *F. australiensis* as a species has been doubted by some authors, but, in common with Lacourt (1968), I regard it as valid here on the basis of the shape of the sessoblasts and the presence of an annulus. Lacourt (1968) states that the two species may be further distinguished by the wider zooids and greater number of tentacles of *F. australiensis*. However, too little is known of these features to allow reliable identification in the absence of sessoblasts.

The zooids, which are usually openly branching, are relatively long (up to 30 mm) and narrow, and may be encrusting or partly free of the substratum.

Both species occur on various substrata, usually near the bottom of ponds and lakes, but rarely on emergent plants or in shallow or running water.

F. australiensis has not been recorded from Britain.

F. sultana has been recorded from the following sites; Belvoir Lake, Leicestershire (SK837327); Broad Pool, W. Glamorgan (SS511910); Broomlee Lough, Northumberland; Derwentwater, Cumbria (NY269209); Easdale Tarn, Cumbria (NY3008), Grisedale Tarn, Cumbria (NY351122); Margam, W. Glamorgan (SS8086); Paddington Canal, London; River Ravensbourne, Kent: Thirlmere, Cumbria (NY322132); Wheatfen Broad, Norfolk (TG3207); Windermere, Cumbria; River Yare, Norfolk.

Hyalinella punctata (figs 2J; 7B)

Although the chitinous zooecium of *H. punctata* is soft and translucent, it cannot be confused with the gelatinous species since these do not form branching colonies. The zooids are relatively short and wide, and are often closely packed together.

H. punctata is not common, and is particularly rare in Britain. There are records for Broomlee Lough, Northumberland; Elterwater, Cumbria.

Lophopus crystallinus (figs 2C; 3)

Colonies of *L. crystallinus* are lobed rather than branching, and the lophophores are orientated in one direction. They are never more than 50 cm across, and occur most often on submerged plants.

L. crystallinus is rare in Britain. River Ravensbourne, Kent; Wheatfen Broad, Norfolk; River Yare, Norfolk.

Pectinatella magnifica (figs 2D; 4)

Pectinatella magnifica forms very large colonies, which are always composed of distinct rosettes of zooids (fig. 4B), and may be found attached to various substrata or floating freely in the water. It does not thrive in running water or at temperatures below 20°C (Marcus, 1940).

P. magnifica has not been recorded from Britain.

Plumatella casmiana (figs 2G; 9B, C)

The colonies are usually entirely encrusting, with closely packed zooids (fig. 9B), although Bushnell & Wood (1971) have reported a single occurrence in North America of 'honeycomb colonies' which resemble those of *P. fungosa*. The zooids usually have a dorsal keel on the zooecium. Early descriptions mention only the commoner thin-walled type of floatoblasts, and either type may occur without the other in a particular colony.

P. casmiana has not been recorded from Britain.

Plumatella coralloides (figs 2L; 8B)

Although for many years regarded as a variety of *P. repens* (Mundy & Thorpe, in press), the bushy colonies of this species are very distinctive. Its apparent rarity may be a result of mis-identification.

Horsham, W. Sussex; Market Drayton, Shropshire; Millwood, W. Glamorgan (SS4888); Renfrew, Strathclyde; Torquay, Devon.

Plumatella emarginata (figs 2H; I; 9A)

The emargination of the orifice from which the name of this species derives is not always present and is in any case difficult to discern. The zooids usually have a dorsal keel, and the colonies are entirely encrusting (fig. 9A). The floatoblasts are often rather flattened at the poles, giving them an almost rectangular outline.

Lliedi Reservoirs, Dyfed (SN5103); Millwood, W. Glamorgan (SS4888).

Plumatella fruticosa (figs 2F; 6A)

Colonies of *P. fruticosa* are almost always erect, attached only at one or two points, and this may be why the species is rare in running water. The zooids are triangular in cross-section, and the zooecium is rather thin and brittle.

Easdale Tarn, Cumbria (NY3008); Grasmere, Cumbria (NY343061); Margam, W. Glamorgan (SS8086); River Rothay, Cumbria (NY347064); Wheatfen Broad, Norfolk; Windermere, Cumbria.

Plumatella fungosa (figs 2L; 11)

This species may form large colonies (a few metres across), and is usually found on solid substrata such as pilings or tree trunks.

Bosherston Pools, Dyfed (SR9794); Crosemere, Shropshire (SJ433306); Dunvant Brickpond, W. Glamorgan (SS596943); Esthwaite Water, Cumbria (NY363968); Grasmere, Cumbria (NY343061); Margam, W. Glamorgan (SS8086); Millwood, W. Glamorgan (SS4888); Newtonmere, Shropshire (SJ426343); Oxwich, W. Glamorgan (SS5087); St. Agnes, Scilly Isles (SV878086); Tresco, Scilly Isles (SV8914); Wrayesbury Reservoir, Berkshire (TQ0274).

Plumatella repens (figs 2K; 10)

P. repens is probably the most common species, occurring on most substrata, especially water plants.

Belvoir Lake, Leicestershire (SK837327); Bosherston Pools, Dyfed (SR9794); Esthwaite Water, Cumbria (NY363968); Fairwood Lake, W. Glamorgan (SS583917); Margam, W. Glamorgan (SS8086); Orielton Lily Pond, Dyfed (SR956992); Tresco, Scilly Isles (SV8914).

Paludicella articulata (fig. 6A)

The colonies have a more delicate appearance than the dendritic phylactolaemates (fig. 6A cf. fig. 10A), and the branching pattern is more regular. *P. articulata* is found mainly in upland lakes, on submerged stones or wood.

Existing records are confined largely to the Lake District, where it occurs at most sites.

Victorella pavida (fig. 6B)

This is the only bryozoan which, although sometimes found in freshwater, can tolerate brackish and marine conditions. The presence of stolons immediately distinguishes it from the other freshwater bryozoans. It commonly grows in docks, harbours and reed beds, attached to a variety of substrata.

V. pavida is known to occur in several European harbours (Prenant & Bobin, 1956), but in Britain may be restricted to the type locality – the Victoria Docks in London (Hurrell, 1927; Prenant & Bobin, 1956).

ACKNOWLEDGEMENTS

I would like to thank Dr P. E. King, Dr J. M. Elliott, Dr G. Fryer and Mr J. E. M. Horne for reading and commenting on the manuscript, and Miss P. L. Cook and Mr P. J. Chimonides for providing specimens from the British Museum. I am grateful to many people and institutions for allowing access to their property, and to the Natural Environment Research Council for financial support.

REFERENCES

Allman, G. J. (1844). Synopsis of the genera and species of zoophytes inhabiting the freshwaters of Ireland. *Ann. Mag. nat. Hist.* **13,** 328-331.

Allman, G. J. (1850). On the present state of our knowledge of the freshwater Polyzoa. *Rep. Br. Ass. Advmt Sci.* **20,** 305-337.

Allman, G. J. (1856). A monograph of the freshwater Polyzoa, including all the known species, both British and foreign. *Ray Soc. Publs*, **28,** 1-119.

Allman, G. J. (1879). Recent progress in our knowledge of the structure and development of the phylactolaematous Polyzoa. Presidential anniversary address. *J. Linn. Soc. (Zool.)* **14,** 489-505.

Blumenbach, J. F. (1779). *Handbuch der Naturgeschichte*. Göttingen.

Braem, P. (1890). Untersuchungen über die Bryozoen des süssen Wassers. *Biblthca Zool.* **2,** 1-134.

Brien, P. (1953). Etude sur les phylactolémates. *Annls Soc. r. zool. Belg.* **84,** 301-444.

Brien, P. (1960). Classe des bryozoaires. In: *Traité de Zoologie* (ed. P. Grassé) **5,** 2, 1054-1335. Paris. Masson.

Bushnell, J. H. (1965a). On the taxonomy and distribution of freshwater Ectoprocta in Michigan. Part I. *Trans. Am. microsc. Soc.* **84,** 231-244.

Bushnell, J. H. (1965b). On the taxonomy and distribution of freshwater Ectoprocta in Michigan. Part II. *Trans. Am. microsc. Soc.* **84,** 339-358.

Bushnell, J. H. (1965c). On the taxonomy and distribution of freshwater Ectoprocta in Michigan. Part III. *Trans. Am. microsc. Soc.* **84,** 529-548.

Bushnell, J. H. (1966). Environmental relations of Michigan Ectoprocta, and dynamics of natural populations of *Plumatella repens*. *Ecol. Monogr.* **36,** 95-123.

Bushnell, J. H. (1968). Aspects of architecture, ecology and zoogeography of freshwater Ectoprocta (Bryozoa). *Atti. Soc. ital. Sci. nat.* **108,** 129-151.

Bushnell, J. H. (1974). Bryozoans (Ectoprocta). In: *Pollution ecology of freshwater invertebrates* (eds C. W. Hart & L. H. Fuller). New York. Academic Press.

Bushnell J. H. & Wood, T. S. (1971). Honeycomb colonies of *Plumatella casmiana* Oka (Ectoprocta: Phylactolaemata). *Trans. Am. microsc. Soc.* **90,** 229-231.

Busk, G. (1852). *Catalogue of marine Polyzoa in the collection of the British Museum.* **1,** 1-54. London.

Cori, C. J. (1941). Ordnung der Tentaculata: Bryozoa. In: *Handbuch der Zoologie* (eds W. Kukenthal & T. Krumbach) **3,** 2, 263-502.

Cuvier, G. (1798). *Tableau élémentaire de l'histoire naturelle des animaux.* Paris. 710 pp.

Dumortier, B. C. (1835). Réchèrches sur l'anatomie et la physiologie des Polypiers composés d'eau douce. *Bull. Acad. r. Sci. Bruxelles,* **2,** 421-454.

Ehrenberg, C. G. (1831). *Symbolae physicae, seu icones et descriptiones corporum naturalium novorum .... Pars Zoologica. I. Animalia evertebrata, exclusis insectis.* Berlin.

Gervais, P. (1836). Réchèrches sur les polypes d'eau douce. *C. r. hebd. Séanc. Acad. Sci., Paris,* **3,** 796-797.

Gervais, P. (1838). Sur les polypes d'eau douce. *Extraits Procès-Verbaux des Séances de la Société Philomatique.* 1838, 129-130.

Goddard, E. J. (1909). Australian freshwater Polyzoa. Part I. *Proc. Linn. Soc. N.S.W.* **34,** 487-496.

Hancock, A. (1850). On the anatomy of the freshwater Bryozoa, with descriptions of three new species. *Ann. Mag. nat. Hist.* 2, **5,** 173-202.

Harmer, S. F. (1901). Bryozoa in Britain. President's address. *Trans. Norfolk Norwich Nat. Soc.* **7,** 115-137.

Harmer, S. F. (1913). The Polyzoa of waterworks. *Proc. zool. Soc. Lond.,* 1913, 426-457.

Harmer, S. F. (1931). Recent work on Polyzoa. *Proc. Linn. Soc. Lond.* **143,** 113-168.

Hayward, P. J. & Ryland, J. S. (1979). British ascophoran Bryozoans. *Synopses Br. Fauna (N.S.),* No. *14*. London. Academic.

Hurrell, M. E. (1927). The ecology of the fresh-water Polyzoa in East Anglia. *Jl R. microsc. Soc.* **47,** 135-142.

Hyatt, A. (1868). Observations on Polyzoa, suborder Phylactolaemata. *Proc. Essex Inst.* **5,** 97-112, 145-160, 193-232.

Hyman, L. M. (1959). *The invertebrates. V. Smaller coelomate groups.* New York. McGraw-Hill.

Jullien, J. (1885). Monographie des Bryozoaires d'eau douce. *Bull. Soc. zool. France,* **10,** 91-207.

Kraepelin, K. (1887). Die deutschen Süsswasser-Bryozoen. Eine Monographie. *Abh. naturw. Ver. Hamburg,* **10,** 1-168.

Lacourt, A. W. (1968). A monograph of the freshwater Bryozoa – Phylactolaemata. *Zool. Verh., Leiden,* **93,** 1-159.

Lamarck, J. B. de (1816). *Histoire naturelle des Animaux sans vertèbres, présentant les caractères généraux et particuliers de ces Animaux, leur distribution,* etc. **2.** Paris.

Leidy, J. (1851). *On Cristatella magnifica* n. sp. *Proc. Acad. nat. Sci. Philad.* **5,** 265-266.

Linnaeus, C. (1758). *Systema naturae* (ed. 10), **1,** 799-821.

Marcus, E. (1925). Bryozoa. In: *Biologie der Tiere Deutschlands* (ed. P. Schulze). Berlin.

Marcus, E. (1926). Beobachtungen und Versuche an lebenden Süsswasser-bryozoen. *Zool. Jahrb. (Syst.)* **52,** 279-350.

Marcus, E. (1934). Uber *Lophopus crystallinus* (Pall.). *Zool. Jahrb. (Anat.)* **58,** 501-606.

Marcus, E. (1940). Mosdyr (Bryozoa eller Polyzoa). *Danmarks Fauna,* **46,** 1-401.

Mukai, H. (1973). Histological and histochemical studies on the formation of statoblasts of a freshwater Bryozoan, *Pectinatella gelatinosa. J. Morph.* **141,** 411-426.

Mukai, H. (1974). A histochemical study of freshwater bryozoan, *Pectinatella gelatinosa. Ann. Zool. Jap.* **47,** 91-102.

Mundy, S. P. (1979). *Studies on the Phylactolaemata.* Unpublished Ph.D. thesis, Univ. of Wales.

Mundy, S. P. & Thorpe, J. P. (1979). Biochemical genetics and taxonomy in *Plumatella fungosa* and *P. repens* (Bryozoa: Phylactolaemata). *Freshwat. Biol.* **9,** 157-164.

Mundy, S. P. & Thorpe, J. P. (in press). Biochemical genetics and taxonomy in *Plumatella coralloides* and *P. fungosa*, and a key to the British and European Plumatellidae (Bryozoa: Phylactolaemata). *Freshwat. Biol.* **10.**

Oda, S. (1974). *Pectinatella magnifica* occurring in Lake Shoji, Japan. *Proc. Jap. Soc. syst. Zool.* **10,** 31-39.

Oka, A. (1907). Zur Kenntnis der Süsswasser-Bryozoenfauna von Japan. *Annotnes zool. jap.* **6,** 117-123.

Pallas, P. S. (1768). Descriptio Tubulariae fungosae prope Wole dimerum mense Julio 1768 observatae. *Novi Commentarii Academiae Scientiarium Imp. Petropolitanae,* **12,** 565-572.

Prenant, M. & Bobin, G. (1956). Bryozoaires, première partie, Entoproctes, Phylactolèmes, Ctenostomes. *Faune de France,* **60,** 1-398.

Prenant, M. & Bobin, G. (1966). Bryozoaires, deuxième partie, Chilostomes Anasca. *Faune de France,* **68,** 1-643.

Rogick, M. D. (1934). Additions to North American freshwater Bryozoa. *Ohio J. Sci.* **34,** 316-317.

Rogick, M. D. (1935). Studies on freshwater Bryozoa. II. The Bryozoa of Lake Erie. *Trans. Am. microsc. Soc.* **54,** 245-263.

Rogick, M. D. (1959). Bryozoa. In: *Fresh-water biology* (ed. W. T. Edmonson) 2nd ed. 495-507. New York. Wiley.

Ryland, J. S. (1970). *Bryozoans.* London. Hutchinson.

Ryland, J. S. (1976). Physiology and ecology of marine bryozoans. *Adv. mar. Biol.* **14,** 285-443.

Ryland, J. S. & Hayward, P. J. (1977). British anascan Bryozoans. *Synopses Br. Fauna* (N.S.), No. *10*, London. Academic.

Saville Kent, W. (1870). On a new Polyzoan, *Victorella pavida*, from the Victoria docks. *Q. Jl microsc. Sci.*, N.S., **10,** 34-39.

Tajima, I. & Mukai, H. (1975). Electron microscope studies on the statoblast of a freshwater bryozoan *Pectinatella gelatinosa.* I. Vitellogenesis in the yolk cell during statoblast formation. *Zool. Mag., Tokyo,* **84,** 205-216.

Thorpe, J. P. & Mundy, S. P. (in press). Biochemical genetics and taxonomy in *Plumatella emarginata* and *P. repens* (Bryozoa: Phylactolaemata). *Freshwat. Biol.*, **10.**

Toriumi, M. (1941). Studies on freshwater Bryozoa of Japan. I. *Sci. Rep. Tôhoku Univ.* (4), **16,** 193-215.
Toriumi, M. (1952). Taxonomical study on fresh-water Bryozoa. II. *Paludicella articulata* (Ehrenberg). *Sci. Rep. Tôhoku Univ.* (4), **19,** 255-259.
Toriumi, M. (1956). Taxonomical study on freshwater Bryozoa. XVII. General consideration: interspecific relation of described species and phylogenetic consideration. *Sci. Rep. Tôhoku Univ.* (4), **22,** 57-88.
Toriumi, M. (1974). Analysis of intraspecific variation in *Lophopodella carteri* Hyatt from the taxonomical viewpoint. XII. General consideration. *Bull. biol. Stn Asamushi.* **15** (1), 1-14.
Trembley, A. (1744). *Mémoires pour servir à l'histoire d'un genre de Polypes d'eau douce, à bras en forme de cornes.* Leide. 322 pp.
Wesenberg-Lund, C. (1939). *Biologie der Süsswassertiere.* Wien. Springer.
Wiebach, F. (1958). Bryozoa. *Tierwelt Mitteleur.* **1,** 8, 1-56.

INDEX TO SPECIES

Page numbers in **bold** type indicate main key references; those in *italic* type indicate illustrations.